William Alexander

The Rinderpest in Aberdeenshire

the outbreak of 1865 and how it was stamped out

William Alexander

The Rinderpest in Aberdeenshire
the outbreak of 1865 and how it was stamped out

ISBN/EAN: 9783741107863

Manufactured in Europe, USA, Canada, Australia, Japa

Cover: Foto ©berggeist007 / pixelio.de

Manufactured and distributed by brebook publishing software
(www.brebook.com)

William Alexander

The Rinderpest in Aberdeenshire

RINDERPEST

IN

ABERDEENSHIRE:

THE OUTBREAK OF 1865,

AND

HOW IT WAS STAMPED OUT.

ABERDEEN:

PRINTED AT THE FREE PRESS OFFICE.

1882.

PREFATORY NOTE.

THE following pages, it may be right to say, lay no claim to direct
official authority. In their compilation full use has been made of
the Minutes of the Rinderpest Association and its Committees, so
as to ensure accuracy as to the main points ; and the newspaper
reports of the day have also been referred to for what was not em-
braced in the formal Minutes. But in addition to these sources of
information, the Compiler was so far conversant personally with
the general character of the proceedings at the time as to, at least,
enable him to avoid giving other than what he believes to be a
correct impression of the spirit and style in which the operations
of the Association were carried out by its Executive.

RINDERPEST in ABERDEENSHIRE.

THE OUTBREAK OF 1865,

AND

HOW IT WAS STAMPED OUT.

Introductory.

BOUT midsummer of 1865, the tenant-farmers and stockowners generally of Aberdeenshire were alarmed and perplexed in no ordinary measure by reports, which by and by proved to be only too well founded, that the fatal cattle disease known as *rinderpest*, or steppe murrain, had been introduced into this country. The first known outbreak was reported to have occurred amongst dairy cows at Islington, London ; and the rapid spread of the plague into other localities seemed imminent. It was on 29th June that two or three cows out of six bought by Mr. Nicholls, an Islington dairyman, were reported as having been seized by the plague; and by the close of July the disease had so spread that hundreds of animals had died, or been slaughtered. The statement, indeed, that £30,000 worth of stock had by that date been lost by dairymen in the Metropolis alone, was believed to be probably under the truth ; and there had been outbreaks at points in several of the midland counties. What this terrible plague of the bovine race might be like, was the first thought amongst those

directly interested. And people who were more or less
" read up," declared that, though hitherto hardly known in
Britain—at anyrate otherwise than historically—*rinderpest*,
or steppe murrain, had long been familiar to Continental
veterinarians. Inquiry into the history of this most viru-
lent of all cattle diseases showed that it had been known
for a very long period ; according to some accounts, for
two thousand years. At anyrate it appeared to have
devastated the herds of the warlike tribes who over-ran
the Roman Empire during the fourth and fifth centuries.
So early as 810 it travelled into France with the armies
of Charlemagne ; and about the same date was supposed
to have visited England, where it again prevailed in
1225. Spreading westward from the Russian steppes,
where it is indigenous, *rinderpest*, as the records showed,
had committed serious ravages among the herds on the
Continent of Europe, at comparatively short intervals,
during every subsequent century, though it was not till
the year 1713 or 1714 that it once more visited England.
On that occasion, too, the first outbreak occurred at
Islington, and in the month of July. In 1744 an outbreak occurred in Holland, which in two years destroyed
200,000 cattle. At that date *rinderpest* seems to have
spread very far, and with most disastrous results. In
Denmark, in five years—1745-1749—it destroyed 28,000
cattle ; and in some provinces of Sweden only 2 per
cent. of the horned cattle escaped. Its ravages throughout Italy were frightful, 400,000 cattle having been destroyed in Piedmont alone. In April, 1745, it was again .
imported into England, where it prevailed more or less
for twelve years. The loss in some cases seems to have
been very great ; the number of cattle that died in
Nottinghamshire and Lancashire alone in 1747 having

been 40,000, while so late as 1757, Cheshire lost 30,000
in six months. In 1714 the plague had apparently been
confined to London and neighbourhood by the slaughtering of diseased animals, and the same remedy was
brought partially into use during the later attack, it being
stated that in 1748 as many as 80,000 cattle in all were
slaughtered, while nearly double that number died of
rinderpest. The only case on record of *rinderpest* having
visited Scotland was in 1770, when the disease was
brought to Portsoy from Holland with some hay. At
that time several cattle died of it, and others, to the
value of £799 12s. 2d., were destroyed, and its further
spread thereby prevented.

As to the symptoms and precise character of *rinderpest* only the most general impressions at first prevailed.
The most accurate information obtainable described it
as originating in a specific virus, communicable by in-
oculation or contagion, and having, like small-pox and
certain fevers, an incubative stage, varying from two or
three to twenty days after introduction of the virus into
the subject till development of the symptoms took place.
When the disease was more fully understood, it was
officially announced that *rinderpest* might be known by
the following symptoms :—" Great depression of the
vital powers, frequent shivering, staggering gait, cool
extremities, quick and short breathing, drooping head,
reddened eyes, with a discharge from them, and also
from the nostrils, of a mucous nature, raw-looking places
on the inner side of the lips and roof of the mouth,
and diarrhœa, or dysenteric purging."

It is perhaps worth while to remark here that when
plague first appeared in England, it was strongly com-
plained of the farmers and dairymen that they pursued

the ruinous policy of keeping back information, and seeking to conceal evidence of the existence of the disease, and the extent to which they had suffered : a policy to which the behaviour of the main body of farmers in the north, as we shall see, formed a distinct contrast.

But, if the terrible *rinderpest* was really threatening or had already invaded the county which, of all others whether in England or Scotland, had made the most marked advance in the breeding of improved cottle of two leading types ; the county whose farmers could claim to have out-distanced others in the production of prime bullocks for the London market ; where farm rents were mainly paid by the sale of fat cattle—the crisis was clearly of the most serious character. And if the spread of plague in Aberdeenshire had to be faced as a practical question, prompt and adequate action on the part of those directly interested was imperative. And that it must be so faced was not long left a matter of doubt.

The first outbreak occurred at Bogenjohn in the parish of Strichen, the disease having been brought thither by a calf received from London about the 20th June, through the medium of a local dealer. The first animal died on 1st July, and betwixt that and 4th August twenty-one had died. The second centre of disease was Westerton, parish of Forgue. The means of infection there were exactly the same as in the other case, viz., a calf brought from London ; and, though the first death did not occur till July 25th, other eighteen had died by 17th August. In each case, it may be said, the whole of the remaining cattle of the herd were removed and slaughtered.

Such, then, was the situation as it presented itself.

It need hardly be said that the reports, which got into rapid circulation, of the outbreak first, and next of the steady progress of the plague at the centres mentioned, were well fitted to alarm the farmers of the county generally, and to stimulate all interested in agriculture to seek out and apply the best means of warding off, or at least coping in the best available manner with the impending calamity ; and it could not be otherwise viewed than as a calamity of a very grave character.

As the subsequent action of the tenant-farmers and landowners of the county in the formation of the "Aberdeenshire Rinderpest Association," and systematic enforcement of the policy of "stamping out" cattle plague, by promptly slaughtering affected animals, on the principle of remunerating such of their owners as conformed to rules framed for the general guidance according to a fixed scale, attracted general attention at the time, and admittedly formed a marked example of successful local organisation and administration on a purely voluntary basis, it will not, it is believed, be without interest, after the lapse of fifteen years, to narrate briefly the course of events, and describe the main points in the procedure adopted during the prevalence of Cattle Plague in the county.

I.—*Initiatory Proceedings.*

The subject of *rinderpest* was first formally taken up at a meeting of the Sub-Committee of the Royal Northern Agricultural Society held on 11th August,

1865. The meeting had been called by the secretaries, Messrs J. & R. Ligertwood, advocates, in consequence of a communication made to them by Mr. M'Combie, Tilly-four, directing attention to the disease which had broken out amongst cattle in London, and pointing to the pro-priety of considering whether steps should be adopted for mutual protection. The meeting was fully attended. And after due discussion, during which the opinion was freely expressed that whatever Government might be induced to do in the exigency that had arisen, the farmers must mainly trust to protecting themselves, it was resolved that, "looking to the grave nature of the emergency, and the cases already reported as having occurred in this county," the Lord-Lieutenant should be asked to call a public meeting of landowners, farmers, and all others interested, "for the purpose of considering the best means of prevention." Meanwhile, it was agreed that two members of committee should proceed to each of the two places in the county where disease had been reported to exist, to make inquiry and report to a future meeting of the committee; the gentlemen appointed being Messrs. Copland, Mill of Ardlethen; and Campbell, Blairton, who were asked to visit the centres of disease in Buchan; and Messrs. Reid, Grey-stone, and Campbell, Kinellar, who were asked to visit the centres in Forgue. The gentlemen so appointed were at once "authorised to employ veterinary sur-geons, and do all things necessary at the expense of the Society." The secretaries were also instructed "to procure, as far as possible, information from the railway and steam companies of the cattle imported into the county during the next week, and report to next meeting of committee."

In accordance with the first resolution adopted by the sub-committee of the Royal Northern Agricultural Society, a public meeting, under the presidency of the Earl of Kintore, as Lord-Lieutenant of the County, was held on 18th August. At that meeting a series of four resolutions was passed. The first resolution, moved by Sir James D. H. Elphinstone, Bart., of Logie-Elphinstone, and seconded by Mr. Copland, Mill of Ardlethen, was as follows :—

"That application be made to the Secretary of State, to extend to Scotland, without delay, the Orders and Regulations of the Privy Council for the prevention of the spread of the cattle plague in this country ; to urge on Her Majesty's Government the propriety of taking measures to prevent the importation of cattle from those countries where the disease is known to exist ; and to procure and publish from time to time such information in regard to the disease as may be obtained from the consuls at the various foreign cattle shipping ports, and, in particular, the Port of Revel."

The second resolution was moved by Mr. Leslie of Warthill, M.P. for the county, and seconded by Mr. George Cruickshank, Comisty, Forgue :—

" That this meeting, in order at once to meet the present emergency, and the loss to parties who have suffered or may suffer from the loss of cattle, by the disease of *rinderpest* in this county, recommend a voluntary subscription for the current year of 1d. per £ on proprietors, and 1d. per £ on farmers, to

be applied at the sight and in the discretion of the
Central Committee appointed by this meeting;
and the meeting recommend the formation of a
Mutual Cattle Assurance Association ; and that any
balance remaining over of the voluntary subscription shall be handed over to the Mutual Cattle
Assurance Society, if such should be established
for this county, in order to form a guarantee fund."

On the motion of Mr. Gordon of Fyvie, seconded by
Mr. Milne of Kinaldie, a large Central Committee was
nominated to carry out the resolutions, and with power
" to arrange for the appointment of sub-committees of
vigilance and protection in each parish or district." And
on the motion of Major Innes of Learney, seconded by
Mr. J. W. Barclay, Auchlossan, the Central Committee
were instructed to communicate with the various district
agricultural associations, in order to their carrying out
the duties of sub-committees in their several districts.

It is not out of place here to say that the first two of
the four resolutions so promptly carried went far to shape
the whole future course and policy of the Association
which sprang out of the proceedings taken, with the
single exception, if it may be so called, that no Mutual
Assurance Society for the county was formed. The
part performed by the Government, as represented by
the Privy Council, was simply to frame Orders regulating
the importation and movement of cattle, which, in no
small part, as will appear further on, was done in accordance with suggestions emanating from Aberdeenshire.
Though the question of compensation from the Imperial
Exchequer for cattle slaughtered was more or less talked
of, that principle was not conceded until after *rinderpest*

had been "stamped out" in the county; and the tenant-
farmers and landholders successfully met the whole requirements of the case by the fixed voluntary subscription
specified in the second resolution. The machinery provided under Resolutions III. and IV., when perfected
under the formal Constitution, also worked most efficiently throughout.

The Central Committee, which met on same day,
directly after the public meeting, and under the presidency of Mr. Gordon of Fyvie, resolved that the
members resident in Aberdeen should form a sub-
committee for consultation as to convening meetings,
framing necessary rules in accordance with the resolutions passed, and for general preparation of business ;
and Mr. J. W. Barclay was appointed convener. The
Secretary was requested to send circulars, with copy of
the resolutions passed at the public meeting, to every
parish minister in the county, requesting him to call a
meeting of the farmers in his parish on an early day in
the ensuing week, with a view to their appointing two
or more of their number to correspond with the Central
Committee. The Secretary was further instructed to
send a corresponding circular to each of the secretaries
of the various district agricultural societies, inviting
him to assist in getting corresponding members appointed
in his parish, and a district committee formed, so as to
secure simultaneous and instant action throughout the
county. As indicating the promptitude with which steps
were taken, it may be stated that circulars bearing date
18th August—the day on which the public meeting had
been held—were within a few hours issued to all the
parish ministers in the county, and to the secretaries of
eleven district agricultural societies. And at the meet-

ing of the Central Committee a week thereafter, it was reported that copy minutes had been received from about fifty parishes in which meetings had been held in response to the circular issued. At same meeting some discussion took place as to the propriety of forming a Mutual Cattle Assurance Association, as had been resolved at the public meeting ; but, the chairman having ruled that it was not competent for the committee to go back on the resolutions, the points raised were left over.

A report by Mr. Morris, V.S., was read to the meeting of 25th August by Mr. Barclay, convener of the Acting Committee, which gave particulars of visits to two farms in the vicinity of the town where cattle had recently died ; and further stated that twenty-seven dairies in and around Aberdeen had been visited. In several cases it had been found that animals had been affected by pleuro-pneumonia, or murrain, but only in one case, that of a calf found in a byre in Wales Street, were there symptoms indicative of the presence of *rinderpest.*

It was at this meeting that the draft "Constitution, Rules, and Regulations of the Aberdeen Rinderpest Association" were first submitted for approval.

II.—*Constitution and Organisation of the Rinderpest Association.*

At the outset of the proceedings, which commenced on 11th August, the duties of the secretaryship, as indicated, had devolved upon Messrs. J. & R. Ligertwood,

secretaries of the Royal Northern Agricultural Society.
These gentlemen, while expressing themselves very will-
ing to give what assistance lay in their power, had intimated that, in justice to their own business, they could
not continue their services in that capacity, and that to
facilitate a satisfactory arrangement they had resolved to
resign their official connection with the Royal Northern
Agricultural Society.   In view of this resignation the
joint sub-committee of that Society and of the Aberdeenshire Rinderpest Association, to whom the matter had
been remitted by the Central Committee, unanimously
resolved to request Mr. J. W. Barclay to act as Honorary
Secretary to the Association, with the assistance of a
clerk, Messrs. Ligertwood acting as Honorary Treasurers.
This arrangement took effect from 1st September, the
date of meeting at which the resolution mentioned was
agreed to.   Previous to this, however, and indeed from
the outset of the proceedings, it had fallen to Mr. Barclay to take a very active part in maturing business, and
carrying out the practical measures adopted.   The draft
Constitution and Rules and Regulations submitted to
the Central Committee on 25th August, were drawn up
by him, and at once commended themselves to the meeting as fully covering the requirements of the case.   With
some verbal alterations, not affecting the principles laid
down, the draft, after full consideration, was formally
adopted as embodying the Constitution of the Aberdeenshire Rinderpest Association.*   The executive agency
under it consisted of the Central and Parish Committees,
county and parish Inspectors, and the Honorary Secretary, whose office, in consequence of the frequent neces-

---

* For "Constitution, Rules, and Regulations of the Aberdeenshire
Rinderpest Association," in full, see Appendix A.

sity for prompt and decisive action, came to be a very onerous and responsible one.

On 26th August, an amended Order in Council had been issued, extending the powers of existing Orders under the Act of 1850 for preventing the spread of infectious or contagious diseases among sheep, cattle, or other animals, to all parts of Great Britain. Under it, authority was given to Provosts and Justices of the Peace in Scotland, to appoint qualified inspectors, with definite powers to carry into effect, within the districts assigned to them, the rules and regulations made under the new and previously existing Orders. At a meeting of the Acting Committee, on the 29th August, the amended Order, which prohibited the movement of animals affected by disease, under a penalty of £20, and which had already been published in the local newspapers, was submitted ; and it was agreed to request the services of the Town and County Police to assist the inspectors (of whom only one, viz., Mr. John Morris, V.S., had at that date been appointed), in the discharge of their duties. And it may here be said that the services of the Aberdeenshire County Police, in particular, intelligently and zealously directed as they were by the Chief Constable, Major Ross, came to be of very great value during the prevalence of *rinderpest* in the county.

At the meeting of the Central Committee, held on 1st September, it was reported that copies of the Rules and Regulations had, along with other papers, been issued to the Parish Committees. And with a view to having the objects of the Association carried out efficiently, the following recommendations were adopted, viz. :—

1. " The appointment of a treasurer, who would also be convener of the City Committee, and the opening of accounts current with each of the three local banks. These accounts to be in the name of the Association, and to be operated on by the treasurer and secretary after being duly authorised by the Central Committee.

2. " The election of six veterinary surgeons as inspectors of the county to be recommended to the Justices for appointment.

3. " That the existence of disease in stock should be certified by one of the county inspectors, as well as by a parish inspector.

4. " That a circular should be sent to parish conveners detailing their duties as per draft submitted.

5. " That the subscription from parties not otherwise liable to the assessment should not be less than £1, and that public companies should be asked to subscribe.

6. " That proprietors, if they see fit, may restrict their assessment to their agricultural rental only, excluding house property, shootings, and fishings.

7. " That if an animal is killed by order of two inspectors, who believe it is labouring under *rinderpest*, the owner ought to be remunerated, as if the disease had been *rinderpest*, although a *post-mortem* examination showed the fact to be otherwise ; but every inspector, on killing the first animal, must verify the accuracy of his belief, by *post-mortem* examination, before proceeding farther."

Along with these recommendations, a general instruction was agreed to be issued to parish committees in these terms :—

" When the owner of cattle observes them attacked

with any disease which he does not know, or suspects to be *rinderpest,* he ought at once to send for the parish Inspector. Should the Inspector be of opinion that the disease is *rinderpest,* the owner of the diseased animals ought immediately to inform the convener of the Parish Committee. The convener of the Parish Committee will immediately request one of the county Inspectors to examine the diseased animals. Should the county Inspector concur with the parish Inspector in believing the disease to be *rinderpest,* he ought to take the measures authorised by the Orders in Council, of which each inspector ought to provide himself with a copy, and the Parish Committee will instantly inform the secretary of the Central Committee, and make a detailed valuation both of the animals diseased and of others which may have been in contact with them. Should the parish Inspector be also a county Inspector, it is desirable that another Inspector, either parish or county, be called in that the disease may be certified by two Inspectors. Inspectors ought to satisfy themselves of the nature of the disease by *post-mortem* examination on the first opportunity, and should they order any animals to be slaughtered, a *post-mortem* examination ought to be made of the animal first killed before proceeding farther."

The following six veterinary surgeons were recommended for appointment, and at once formally appointed county inspectors by authority of Justices of the Peace :— Messrs. John Morris (previously acting), Aberdeen ; Robert Sorley, Alford ; Stewart, Rothiemay ; Thomas Hay, Ellon ; M'Gilvray, Rayne ; and Dewar, Midmar.

At the meeting just referred to the City Committee in their report, begged " to draw attention to the cure by chalybeate water, or water off rusty iron, recommended by the British Consul at Warsaw, published in all the newspapers. The cure is so simple that it would be injudicious," they say, "not to use it as a precautionary measure."

The statement of the Consul here referred to was to the effect that on asking information from several large landed proprietors, he had been informed that, when cattle plague was prevalent in Poland in the year 1857, "no remedy answered so well as, on the first appearance of the malady, to give the cattle water very strongly impregnated with iron. This treatment," it was said, "was first discovered from its having been observed that upon a farm where there was a chalybeate spring, the cattle were either but slightly affected or recovered while drinking profusely of the chalybeate water, while on the adjoining farms the beasts died in large numbers. It was then found that by putting old iron into the cattle troughs, so as to produce a highly chalybeate water, the same result ensued, and the cattle recovered." Experience soon taught the committee and all concerned that "simple" cures by mineral water and the like would be of little enough avail !

Up to this date, and even later, a controversy prevailed, not only as to the treatment of the disease, but even as to its origin, and the expediency of imposing any restrictions whatever on importation and inland transit of cattle. The theory of spontaneous origin of the disease, by filthy and crowded housing and the like, was pertinaciously insisted upon by not a few who scouted the idea of interference with perfect liberty as regards the admission of foreign stock as an infringement of the principle of free trade. And it was in reference to the general position of matters indicated that an eminent French veterinarian, Professor Bouley, uttered himself in these words :—" If England had not been disarmed by her laws it had been possible to have arrested its [the plague's] march by energetic measures like those adopted on the Continent, where the Austrian and Prussian sentinels, always vigilant, guard Western Europe, and protect it against the invasion of this Russian scourge. But in England, as it appears, the bovine race must perish rather than a principle. The maxim is sublime, no doubt, but it will be productive of great disasters." Fortunately, the extreme doctrinaire notions referred to found but little countenance in Aberdeenshire.

By 26th August sixty-nine parishes out of 85 within the county had reported to the Central Committee, and agreed to the Association's resolutions and rules, with slight suggested alterations. And the meetings of the Committee thenceforward were held weekly, on Fridays, up to February, 1866.

III.—Administration.

At the meeting of the Central Committee on 8th September, reports were given in of cases of *rinderpest* among 24 cattle in a field at Craigiebuckler, belonging to Thomas Gordon, cattle-dealer (and which it was believed had infected cattle in an adjacent field) ; at Garmond ; at Middle Barnyards, Peterhead ; and at Strichen Mains ; also at Westerton, Forgue, the latter being a renewed outbreak. At Garmond it appeared that ten animals, several of them calves, had died by *rinderpest*, and two, a cow and a calf, were slaughtered by order of the Inspectors. Three cattle had died at Middle Barnyards. At Strichen Mains twelve in all had been attacked, nine being calves ; and of the whole—nine died, two were slaughtered, and only one (a calf) recovered. In this case it was reported that the cattle were in a field together, and were not known to have been in contact with other animals. In the case of the cattle at Craigiebuckler, the circumstances were different. The owner, Thomas Gordon, was a dealer, and constantly passing animals through his hands. The two first attacked, after being separated from the rest (some twenty-four in all) by the Inspector's instructions, were, against his orders, removed by Gordon, who would give no account of them ; and there was the further aggravation that Gordon, at the imminent risk of communicating plague to the cattle belonging to a respectable farmer in an immediately adjoining field, had acted against the Inspector's instructions in removing cows into the Craigiebuckler field. The remaining cattle there, belonging to him, were, after re-inspection, slaughtered and buried " three feet deep " in the earth, by order of the Executive Committee of Inspectors.

The Acting Committee and Inspectors had now got fairly into their work ; and nothing will better indicate the style in which that work was carried through than a few extracts from the reports of Inspectors, and from the resolutions adopted, with summaries of general proceedings, presented as nearly as possible in the actual terms of the reports given in.

At the Central Committee meeting of 15th September, a report by Mr. Hay, V.S., Ellon, to the Hon. Secre-

3

tary was read, stating that on the 9th of the current
month he had visited Kirkton of St. Fergus, where he
found three calves in the last stage of *rinderpest*. He
adds—

" I destroyed the three, the *modus operandi* being by
shooting them. I then went to Mr. Milne, Barnyards,
along with Mr. Laing, and found (in addition to the
three cows and one calf which I saw on the 7th in
various stages of the disease, and which were shot on the
8th) another cow suffering from *rinderpest*. I shot
her, and of all Mr. Milne's cattle that have been
in contact with the affected animals, only two remain,
and they are as yet sound."

On 9th September, two calves had been slaughtered
at Strichen Mains, " being the last out of two cows, one
quey, and nine calves (except one which got better), that
were in a park together," and not known to be in contact with other animals. Then from Messrs. Stewart
and Snowball, V.S., comes this report—

" This is to certify that a cow belonging to Mrs.
Booth, Westerton, Forgue, died of *rinderpest* on the
morning of the 6th inst., and was buried with the skin.
To-day we found a cow and two calves ill of *rinderpest*,
and have caused them to be destroyed in our presence,
and buried with their skins. The above are the whole
of Mrs. Booth's stock, and her houses will be disinfected
as soon as possible."

Again, at the meeting of 22nd September, Messrs.
Stewart and Barron, V.S., report having ordered the

slaughter of a cow and calf at Garmond. And Mr. Hay, V.S., reports—

"At the request of Mr. Logan, convener of the St. Fergus Parish Committee, I went on Saturday last to the village of Ugie, and saw a cow belonging to Mr. Lily, in the last stage of *rinderpest*. I destroyed her by shooting her. Mr. Lily has other two cows, two calves, and two one-year olds, which have been in contact with the diseased cow. I next went to Mr. Hastie, Stonemill, St. Fergus, where the convener—Mr. Logan—the inspector, Mr. Elrick, and Messrs. Watson and Penny, both members of the committee, were waiting for me. This is the first time I have met members of committee on the ground, and I must say they appear to be men of energy, and determined to wrestle with the foe that has for the first time made its appearance in the parish ; for, although the hour (6·30 p.m.) when I left them was rather late, they were determined to get six fat oxen that had been in contact with *rinderpest* killed and dressed that night if possible. I found a two-year-old ox in a far advanced stage of *rinderpest*, and a calf showing evident symptoms of the disease. The ox I destroyed, and made a *post-mortem* examination in presence of the Inspector, &c., as neither of them had ever seen a case of *rinderpest*, and I am perfectly satisfied of the disease being *rinderpest*. Six two-year-olds and seven calves have been in contact. I next went to Mr. Stewart, Cairnhill, St. Fergus, and found a two-year-old quey also in a far advanced stage of *rinderpest*. I ordered her to be destroyed. She has been in contact with twenty-two animals. This is the most inexplicable outbreak I have met with. In fact, there is no accounting for it either by infection or contagion, and I do not believe in its spontaneous origin."

A couple of days after two other cows were ordered to be slaughtered at the same place.

The question of treatment of cattle that had exhibited symptoms of plague was still a matter of discussion in many quarters, and at the meeting of the Central Committee, on 29th September, in order to test the practical view of the question in light of the experience had, the veterinary surgeons present were invited to give their opinions as to whether it would be advisable to attempt treatment of any diseased animals in place of slaughtering. Three of the Inspectors—Messrs. Sorley, Hay, and Stewart, the latter by letter—declared in favour of prompt slaughtering where disease was apparent ; while one—Mr. Dewar—thought he should like to try treatment of the animals. Following upon the opinions expressed, the meeting promptly and unanimously resolved—

" That this Committee, having heard reports from several county Inspectors as to the nature of *rinderpest* and its highly contagious character, and having carefully deliberated on the whole question, are unanimously of opinion and hereby direct, as the best means of getting the disease extirpated and of lessening the loss of all parties interested—
1. That all animals diseased, or which have been in contact with disease, be slaughtered with the greatest possible despatch ; that till this is done and the premises disinfected watchers be stationed at every centre of disease to prevent its spreading

either by persons or animals ; and that all animals
within one mile of any centre of disease be immediately housed and kept so.   2. Any member failing
to attend to these directions shall forfeit all claim
on the funds of the Association.   3. It is remitted
to Parish Committees to see that these directions
are carefully attended to."

Relative to this resolution, specific instructions to
Inspectors and for disinfection of premises, in accordance
with the then newly-issued Orders in Council, were
adopted.   It was set forth as the duty of every Inspector
to make himself thoroughly acquainted with the Orders
in Council ; where disease was reported the Inspectors
were instructed to visit the place at once, whether sent
for or not, and to endeavour to trace out how it had
been communicated, especially in the case of new centres ;
and where the slaughter of any animal had been ordered,
they were specially instructed to make a *post-mortem*
examination, reporting the symptoms prior to death, and
as brought out by the *post-mortem* examination, to the
Honorary Secretary.   In the instruction bearing on the
general treatment of affected animals, it was said—

"The Central Committee are of opinion that the
    best means to get the disease extirpated is to
    slaughter all animals diseased, or which have been
    in contact with disease ; but as the Orders in Council
    only authorise the slaughter of *diseased* animals,
    Inspectors cannot, without consent of their owner,
    slaughter healthy animals which have been in con-
    tact with disease.   If the animal is in good condi-
    tion, and the disease not so far advanced as to

injure the meat, it will be advisable to slaughter
and dress the animal.   If fit for use, the owner
will take charge of carcase; but the Inspector is
bound to see all the offal, blood, bedding, &c.,
properly buried, and the skin and tallow carefully
disinfected, &c. And if the animal is in an advanced
stage of disease, it only remains for the Inspector
to see it killed, and buried five feet deep, as directed
by the Orders in Council."

As will be noted, the Committee went in advance of
the Orders in Council by expressing a distinct opinion
in favour of slaughtering animals which had "been in
contact with disease," as well as those actually diseased.
And, as a further measure of precaution, the Honorary
Secretary was instructed to represent to the Magistrates
of Burghs and Justices of Peace for the County, as the
proper authorities, "that this Committee is strongly of
opinion, and anxiously wish, that all fairs and markets
be stopped for two months, from October 1st, so far as
regards cattle and sheep; and to request them to
exercise their powers under the new Orders in Council
to this effect."

At the meeting of 6th October, it was reported that
£3200 of the voluntary assessment had been received;
and in reference to the stoppage of fairs and markets, a
deputation from the Aberdeen Flesher Trade, requesting
the continuance of the King Street Market for the sale
exclusively of fat cattle, was heard. The members of
the deputation stated that their constituents were quite
agreeable to having the county markets shut up; but
considered that it would be safe, as well as convenient
for all parties, if fat stock were allowed to be exposed in

King Street. The difficulty of strictly defining "fat
stock" was pointed out, and the consequent risk of
having the market "inundated with cattle from all
quarters," especially from southward, dwelt upon. The
alternatives seemed to be opening the market for
animals strictly consigned for slaughter, or adopting a
suggestion made to allow markets confined to single
parishes to be held. The Honorary Secretary having
pointed out that the Order in Council gave no discretionary power in the direction of the suggestion referred to,
it was agreed to allow consideration of the subject to
lie over for a month.

At this meeting the question of sales of cattle
"warranted free of all disease," by jobbers, was brought
up by Mr. M'Combie, Tillyfour. A "jobber," in or
near his district, who had advertised seventy-four cattle
for sale, had, he stated, set at nought all his remon-
strances as Chairman of the Central Committee. And
as he brought his cattle from distant centres that
were in communication with Falkirk, the risk of disease
being thus imported must be great. "In my district," said the speaker, "we are surrounded with a
host of jobbers ; and even farmers, who have very little,
some of them, to lose, will go on, and are daily going
on, with their jobbing, to the risk and annoyance of
every one in the county who wishes to pay twenty
shillings in the £1. The lung disease has appeared,
if not the *rinderpest*, at my very threshold (within three
miles), brought in by jobbed cattle. I called yesterday,
and the owner reported to me that he had resolved to
kill every cow that he had, two or three being already
dead, and other five nearly all affected. Nothing but
the strong arm of the law will prevent these parties from

going on in their practices ; if you speak to them, they will laugh at you in your face, and tell you you cannot prevent them, and give you abuse to the bargain. Our country is in the most critical position. One penny in the £1 from landlords, and a penny from tenants, will do little to reimburse the farmers in this country if the disease should become general. No one can tell where it may end. What, indeed, have tenants to depend upon but their cattle, and the landlords can have no rent if the cattle die."

On the motion of Mr. M'Combie, seconded by Mr. James Porter, land steward, Monymusk, it was agreed—

" 1. That this Committee memorialise, through the Lord-Lieutenant of the County, the Lords of Her Majesty's Privy Council, praying them to empower Local Authorities when they shall deem it expedient, as a measure for preventing the spread of *rinderpest*, to prohibit sales of cattle by public auction.

" 2. That purchasers of cattle at the public roups of cattle jobbers shall forfeit all claims on the funds of the Association."

A week after the date of the meeting just referred to, the funds of the Association were reported to be about · £3400. Claims for compensation had, at this date (13th October), been made to the extent of about £600, and it was thought the total amount of the claims accruing was not likely to exceed £2000. The total number of beasts reported as affected with the plague, and dead, or slaughtered, up to date had been 283, of which 174 had been in the parishes of Peterhead, St. Fergus, and Longside. On 20th October another outbreak was reported at Mill of Tiffery, Longside, being the second at same place, when, by recommendation of the Parish Committee, the whole remaining stock, consisting of five cows and four calves, valued at about £109, were slaughtered. At the succeeding weekly meeting, the Honorary Secretary was able to report the " very gratifying information " that no case of *rinderpest* had occurred since previous meeting, and that about a month had elapsed since a new centre had appeared. A donation of £50 to the funds of the Association, from Her Majesty the Queen, was intimated. And a legal opinion having been received from the Crown Office to the effect that periodical sales of cattle fell under the provisions of the Order in Council applicable to stoppage of fairs and markets, the Honorary Secretary was instructed to send copies to the parties who had advertised sales.

The proceedings at the meeting of Central Committee, held on 19th November, furnish a very good example of the thorough-going yet cautious way in which the business of the Association was carried on. Mr. Hay, county Inspector, reported upon a case of *rinderpest* at Middle Hythie. The tenant, who had ceased to be a member of the Association, was found to have just killed three cattle (in addition to nine previously killed), driving the carcases to Peterhead, while the offal was allowed to lie about the farm, merely covered with some straw, &c. Four had been killed, and the carcases in process of being buried in the farmyard dunghill ; and other six had been driven to Peterhead for slaughter. In short, a distinct and deliberate violation of Privy Council Orders had been committed. Mr. Hay, who, on stepping in as Inspector, had destroyed five cattle additional, commented severely upon the recklessness of the farmer; and a correspondence read to the meeting showed that the Honorary Secretary had communicated with the Parish Committee, pointing out that laxity had occurred on their part in not having taken such measures as might prevent risk of contagion; and, in particular, that the parish Inspector ought to have carried out the instructions, under Orders in Council, to slaughter all diseased animals, and to prevent animals being removed from the farm, &c. The meeting resolved to minute that, in their opinion, the Parish Committee, as responsible in the matter, ought to have taken action immediately on hearing the current rumour of disease among the stock; also that they were very much dissatisfied with the conduct of the Inspector; and it was resolved to instruct Messrs. Ligertwood to take the necessary steps to prosecute the owner of the cattle. At a subsequent meeting, after explanations received, it was, however, resolved, and minuted, that the Old Deer Committee, in their action in the Middle Hythie case, had conformed to the general instructions.

The summary report of the completed case at 'the farm mentioned was that "the disease is quite extirpated at Middle Hythie; 28 are dead and buried on the farm; 12 have been slaughtered and sent to Peterhead; 14 have been driven alive to the same place; while four cows have been slaughtered and sent south."

On 11th November, at their usual weekly meeting, the Central Committee had had under consideration resolutions (1) to petition the Privy Council to grant powers to Local Authorities to prohibit cattle being brought into the counties for which they acted "from any place in Great Britain or Ireland, or from particular districts thereof" (cattle in transit by railway excepted) ; and to issue an Order "prohibiting cattle being brought from any place in Great Britain to any place in Scotland north of the river Dee and the county of Perth"; and (2) extending the period during which cattle markets might be prohibited by two months. A statement accompanied the resolutions, which narrated the action taken since the outbreak of *rinderpest* in the beginning of July, showing that 283 cattle had died, or been slaughtered, involving claims on the funds of nearly £2000, while £3750 had been raised by voluntary assessment to meet such claims. It then sets forth that by the action taken the disease had been stopped—"the last death of animals infected having been on 19th October last."

Unfortunately, however, for the county (it was added) a farmer, reckless of the risk he incurred, brought into the county on 11th October a lot of fourteen cattle, the Committee believed, from Falkirk. Knowing the risk of this proceeding, the Committee instructed an Inspector to examine these cattle, which he did on 13th October, and again on 18th, but found them all apparently healthy. The disease was, however, latent in the animals, and, on or before the 1st November, had developed itself among the farmer's whole herd. Notwithstanding the Inspector's prohibition, part of the animals were sent across the county, and already the disease has appeared most virulently at four different places in that neighbourhood.

The Committee expressed themselves as very unwilling to interfere with freedom of trade, but, in justification of stringent action, desired to say "they see that, so long as unprincipled dealers are allowed to bring in cattle from infected districts, their labour is completely in vain, and that they may, for any prospect of ex-tirpating the disease, abandon their exertions in despair." Hence the drift of their draft petition to the Privy Council.

IV.—*The Policy of the Association Tested.*

The state of alarm into which the country had been brought during the autumn of 1865, by the prevalence of *rinderpest*, led to the appointment of a Royal Com-mission on the subject. The Commission reported soon after the middle of November; and at the meeting of the Central Committee of the Rinderpest Association, on the 24th of that month, their report came under consideration, when the Honorary Secretary submitted a statement on the subject. The Commissioners recom-mended (1) the suspension, "for a limited time," of all movement of cattle; or (2) certain detailed pre-cautionary regulations, should the first, and in their view preferable, recommendation not be approved. In his statement, the Honorary Secretary, who throughout founded on the experience had in dealing with *rinderpest* in the county, objected to the first proposed measure (1) as being most impracticable; and (2) even if practic-able, that the loss and inconvenience would be as great as from the disease; and (3) that it would not have the effect expected. On the other hand, the precautionary measures suggested by the Commissioners were deemed

too ccmplicated to be workable ; the Association having
found considerable difficulty in getting even simple and
explicit instructions that had been issued attended
to by the members and officials in carrying out the
"stamping out" process, which had been adopted by the
Association.   The facts, briefly put, were these—*Rinderpest* had broken out at four distinct centres (in each case
by calves brought from England), and cattle had been
affected on thirty-two different farms, the number that
had died or been slaughtered up to date being 367.   At
three of the centres—Strichen, Forgue, and Monquhitter
—the plague had been certainly "stamped out," the last
animal slaughtered at any of these centres having met
its fate so long ago as 16th September.   At the fourth
centre, in Peterhead parish, the disease had spread considerably, and been carried into the neighbouring
parishes of Longside and St. Fergus.   In all these
cases, slaughtering had been resorted to, and the plague
was stopped, although, as a shorter time had elapsed
since the process was completed than in the case of the
first three centres, it was not yet certain that the disease
had been conclusively extirpated.*   The only two points
in which the Association had found its powers inade-
quate were (1) in regard to the imposition of an assessment; and (2) the prevention of cattle from coming into
the county.   Had the Association possessed the latter
power in time, it was not doubted that the county would
have been earlier clear of the disease.   The Honorary
Secretary submitted the following proposal for consideration, as covering all that was required in the existing
circumstances :—

    "An Act of Parliament authorising the formation of

* After experience proved that it had not.

County Boards, consisting, in Scotland, of the Commissioners of Supply and tenants elected by each parish, the jurisdiction of this Board to extend over the whole county, royal burghs alone excepted. This Board to be invested with the following powers :—1. Those possessed by. Local Authorities under the Order in Council of 22nd September. 2. To prohibit any cattle from being brought into the county ; if advisable to admit cattle from certain districts only, retaining the general prohibition applicable to all other places not thus specially excepted ; but not to prevent the transit of cattle by railway through the county, under proper precautions. 3. To impose an assessment on the agricultural rental of the county not exceeding in any one year   per pound : one half to be paid by proprietors and the other by tenants, this fund to ·be employed in indemnifying sufferers, paying inspectors, officials, &c. The adoption of the Act by counties to be optional, and to be determined by a public meeting of the county, convened by the Lord-Lieutenant on a suitable requisition, or when the cattle plague breaks out."

The Order of 22nd September here referred to authorised every Inspector within his district, *inter alia,* "to seize and slaughter, or cause to be seized and slaughtered," and to be buried five feet deep, "any animal labouring under" cattle plague ; and to direct animals that he suspects to be labouring under the said disease to be kept separate from animals free from it. And it empowered Local Authorities, on due notice given, to exclude animals (cattle, sheep, goats, or swine) from any specified market or fair within their jurisdiction, for a time to be specified in published notice.

The resolution of the meeting was to refer consideration of the whole matter to a meeting of landed proprietors, occupiers, and others, to be called by the Lord-Lieutenant. And on 1st December, accordingly, a meeting so constituted was held, at which, after considerable discussion, the following resolutions were adopted :—

" That this meeting, in expressing a general approval of the report of the Royal Commission appointed to enquire into the origin and nature of the cattle plague, is of opinion—

" 1. That, to suspend, even for a limited time, the movement of cattle from one place to another in Great Britain would, were it practicable, be highly inexpedient, and likely to be attended with serious consequences, not only to those more immediately interested in the cattle trade, but to the public in general.

" 2. That the suspension of fairs and markets, and public sales, and exhibitions of cattle, ought, under existing circumstances, to be continued for a further period.

" 3. That, with a due regard to the interests of the community at large, no unnecessary impediments should be put in the way of fat or stock cattle being imported, but that all such cattle should be landed only at certain specified ports, and retained in quarantine until slaughtered, and that no lean or store cattle ought to be admitted into Great Britain from any place beyond seas under existing circumstances.

" 4. That the Orders in Council issued from time to

time, investing Local Authorities with certain
discretionary powers, have been only partially successful in arresting the progress of the plague, owing
to the want of uniform action in counties and
districts, and also to the difficulty of enforcing rules
and regulations on officials and interested parties.
Inspectors ought to have the power to prevent
any cattle whatever from being removed alive from
any infected farm ; and that during the time of
disease, and for twenty-one days after the last case ;
also from suspected farms ; but in this case the
prohibition not to exceed ten days, provided disease
does not break out.   Further, that powers ought to
be granted to cause all dogs to be chained, or kept
under immediate control.

" 5. That experience having proved in this — one of
the largest and most important agricultural and
cattle-breeding counties of Scotland—that the most
efficient mode of dealing with such a matter is by
a system of local organisation and local control,
this meeting would regard with satisfaction an Act
of Parliament authorising the formation in counties
of a Central County Committee, to be composed
of landowners and tenant-farmers, with power to
raise a compensation fund ; to proscribe neighbouring and other counties known to be infected with
the plague, and refuse to admit their cattle ; to pro-
claim in the local newspapers infected districts ; and
to adopt, in regard to those districts, as well as in
the county generally, such measures of active interference and precaution as may appear calculated to
meet any emergency that may arise.

" 6. That a copy of the resolutions passed at this meeting

be transmitted by the noble Chairman to the Home Secretary, and to the Clerk of the Privy Council, and the Secretary of the Royal Commission."

At the ordinary meeting of the Central Committee on 8th December, it was unanimously agreed to have the statement of the Honorary Secretary, to which we have referred, embracing, as it did, a "succinct record of the measures adopted in this county for putting down the cattle plague," printed and circulated.*

This resolution was come to after full discussion and consideration at two meetings, a few influential members at first inclining to approval of the recommendation of the Royal Commissioners to suspend all movement of cattle. But latterly the statement and suggestions of the Honorary Secretary, whose executive ability and mastery of the whole elements of the case were strongly spoken to, commanded general assent in the Committee and out of it.

In course of the discussion, reference was made to the destructive progress of the disease in the neighbouring counties to the southward where no effective steps had been taken. In Forfarshire, in particular, it was stated that no fewer than 331 beasts had died in a single week. It was believed that the circulation of the statement, and its being put into the hands of the Lords-Lieutenant and Conveners of all the counties of Scotland, as well as its being submitted to the Privy Council and the Royal Commissioners, would be of great service, as a means of maturing opinion and leading to useful action. The Privy Council had already, it was

* This statement is fully quoted from further on.

4

stated, given considerable weight to the opinions and practice of the County of Aberdeen in the matter; it being known on good authority, that several of their Minutes and Orders had been based pretty much on recommendations sent up from the Association; and no doubt was entertained that due consideration would be given to the views of the Honorary Secretary in any measures that might be framed.

At this point of time the question of "treatment" of animals labouring under *rinderpest* again came up in an incidental way. Certain papers had been received by the Convener of the County from Dr. Lyon Playfair, who, it was reported, wished "to get the treatment recommended by the Edinburgh Commission tried under intelligent direction in Aberdeenshire." This application was disposed of literally in a parenthesis, and in these terms:—"Treatment of cattle being contrary to the policy adopted by this Association, and, happily, no case of disease to treat, the Secretary was instructed to reply to the Convener's note to this effect."

At the meeting of the Central Committee on 15th December, two outbreaks of plague were reported, one at Stranduff, Kincardine O'Neil, the other at Newton of Auchinclech, parish of Skene. In both cases, contagion had been brought by cattle, bought from a dealer, and which had entered the county from Kincardineshire, across the Cairn-'na-Month. In one of the cases, seven cattle out of a stock of twenty-four had died, and the rest were slaughtered; in the other, two out of nine had died, and the rest were slaughtered. Action had been promptly taken in each case by the Conveners of the Parish Committees in which the outbreaks had occurred, along with the Honorary Secretary. It was also reported that, after an interval of twenty days, a fresh outbreak, believed to have originated from insufficiently-buried offal, had occurred at Hythie, Old Deer ; three cattle affected had been slaughtered, and stringent instructions given to the several Parish Committees that all manure and offal heaps should be covered at least a foot deep with earth. The Honorary Secretary further reported that, to meet the danger on the southern boundary, watchers had been stationed along the border line of the county, at all necessary points, to prevent ingress of cattle, and at the various centres of disease, to see that proper precautions were attended to. At the immediately succeeding meeting, it was agreed, after discussion, that "double watchers " should be stationed on all the roads leading into Aberdeenshire from the south, no danger being apprehended on the other sides of the county.

At the meeting of 15th December, in view of the fresh outbreaks reported and dealt with, Mr. Barclay, as Honorary Secretary, was fully empowered to " purchase or order the slaughter of " diseased animals, or animals that had been in contact with disease, belonging to non-members of the Association, as he might find advisable, he being relieved of all responsibility on thereafter reporting his proceedings to the Committee.

The question of additional powers to Local Authorities had been under the consideration of a Sub-Committee, who, on 22nd December, reported to the Central Committee their having laid the following resolutions before a meeting of Justices of the Peace, by whom they had been approved :—

" That the Lords of Her Majesty's Privy Council be solicited :—

" 1. To prohibit all fairs and markets within the United
Kingdom.

" 2. To grant powers to Inspectors to kill diseased
cattle under warrant of two Justices of the Peace.

" 3. To grant powers to Local Authorities to prohibit
raw hides, straw, or fodder from being brought into
their county.

" 4. To order that all foreign cattle be slaughtered at the
port of debarkation.

" 5. To issue some stringent orders against allowing
dogs to stray through the country.

" 6. To prohibit all movement of animals within an area
of one mile from any diseased animal or farmstead
during the existence of disease, and for six weeks
after the last case ; and to grant power to any
Inspector to prohibit for ten days the removal of
any animal from any farm where he suspects
disease to exist.

" 7. Further, to provide that no animals be moved from
any part of the county to another, without a
certificate of health from a special Inspector, and a
licence from a local magistrate."

After considerable discussion, turning mainly on
paragraph 7, these resolutions were carried in the Central
Committee, though only by a small majority. The 7th
resolution was admittedly in the nature of a compromise,
going so far in the direction of the recommendation of
the Royal Commission stopping all movement of cattle ;
and it had been practically adopted by some other
counties. To this it was replied by the Honorary
Secretary, who, along with others, opposed the resolution, that it would create a great deal of incon-

venience and trouble to many parties without yielding
corresponding benefit. As to what had been done in
counties in the southern portion of the Kingdom that
ought not to carry much weight, inasmuch as, when the
County of Aberdeen met originally and resolved to act,
these counties met and resolved to do nothing. The
County of Aberdeen had, the Honorary Secretary held,
carried the day by the course it had taken, and he failed
to see how the people in the south should be right now
any more than when they previously proved themselves
to be wrong.

Such, generally, was the position in which matters
stood at the close of the year. A new Privy Council
Order had been issued on 16th December, which seemed
to give Local Authorities almost absolute control over
the movement of cattle within their jurisdiction. What
the Central Committee, which never conceded the
principle of absolute prohibition of the movement of
cattle, did, was to agree to continue to watch the
southern boundary of the county closely, and to look
after centres of disease, leaving it to individual districts
that might desire more stringent regulations to apply
to the Local Authorities on their own account. A
Notice on the subject was soon thereafter issued by the
Justices of Peace for the County of Aberdeen in virtue
of the powers contained in the Order of 14th December,
and in a subsequent Order dated 20th January, 1866.
It was in these terms :—

"*First*—That it shall not be lawful to remove any
    animal, as defined in said Orders, from any house,
    courtyard, or place within the distance of one mile
    from any animal attacked with Cattle Plague, or to

bring any animal as so defined to any place within
one mile of any animal so attacked ; and that from
the day any such animal is attacked to the first day
of March next, provided always that nothing
contained in this Notice or Declaration shall make
it unlawful to remove sheep from within the
distance of one mile from any animal so attacked
during the three days immediately following the
first attack of any such animal with Cattle Plague,
with the licence of an Inspector and the Convener
of the Parish Committee of the Aberdeen Rinderpest Association acting for the parish in which said
sheep may be at the time.

" *Second*—That it shall not be lawful to remove any
   hide, skin, horn, hoof, offal, dung, hay, straw, fodder,
   or litter from the district within a mile of any
   animal attacked with Cattle Plague, to any other
   place beyond said district, and that for the period
   from the day any animal is so attacked to the first
   day of March next."

The penalty for breach of Order was £20.

A statement of the claims on the Association, submitted to a general meeting through the Finance
Committee on 12th January, 1866, showed that the
income up to that date from voluntary assessment and
donations amounted to £4133 7s. 6d. ; and the liabilities,
including £175 1s. 8d. of expenditure in working
charges, amounted to £2471 16s. 10d.   From the
particulars of claims it appeared that 68 cattle had died,
of the value of £745 5s. ; 111 cattle had been killed and
buried, of the value of £1213 11s. ; 205 had been killed
and the carcases sold, of the value of £3391 13s.   The

proceeds of sales had amounted to £2257 2s. 1od.; and the absolute loss to £1134 1os. 8d.   About £500 of this loss was on cattle that had died before the "stamping out" principle was adopted, and it included payment of full value of first cattle slaughtered at Forgue on the outbreak of the plague.[*]   The total loss amounted to £3006 6s. 8d.   The total claim on the funds at the rate of two-thirds of value of cattle allowed to die, and three-fourths of value of those slaughtered, was £2296 15s. 2d. At date of the accounts, 5th January, the total claim of each parish in which disease had broken out stood thus :—

Strichen ...	...	...	£276	6	7
Forgue ...	...	...	314	0	10
Monquhitter	...	...	104	17	6
Peterhead	...	...	350	7	0
St. Fergus	...	...	306	9	6
Longside	...	...	485	3	0
Lonmay ...	...	...	243	13	5
Old Deer	...	...	63	7	4
Kincardine O'Neil		...	64	10	0
			£2296	15	2

In their report to the general meeting the Central Committee recommended that immediate payment should be made of the claims of those who had suffered by slaughtering; and this was accordingly done.

[*] When plague broke out at Westerton the farmers in the locality undertook to bear the loss incurred by slaughtering Mr. Ogg's cattle, the Association not being then formed.   And to relieve them it was agreed to pay the full amount, £282 15s.

V.—*Concluding Period.*

The general meeting of the Aberdeenshire Rinderpest Association was held on 12th January, 1866, to
receive report and statement of accounts, of which a
summary has been already given. The report stated that
rinderpest had been brought into the County of Aberdeen
by cattle on seven different occasions, and had prevailed,
more or less, for six months. The measures adopted in
each case to cope with the disease appeared to have
been successful. The last case of slaughtering had
occurred on 26th December, and there had been no case
of disease in the county since. Every parish in the
county had joined the Association, and the voluntary
assessment had been remarkably well paid, the deficiency
not exceeding, probably, 15 to 20 per cent. This, it
was stated, was, in a great degree, due to the exertions
of the Parish Committees, who had zealously attended
to their duties. Although the disease seemed for the
present subdued, the Central Committee were fully per-
suaded that it would be unwise to relax, in the slightest,
their vigilance in preventing it from being again brought
in. They, therefore, recommended the members to
prepare, as far as they possibly could, for a lengthened
exclusion of cattle from the south. As soon as
they believed it safe, the Committee would recommend
the Justices of Peace to modify or altogether
rescind the existing Notice prohibiting cattle being
brought in from the southern counties. They considered
that the greatest danger now to be apprehended was the
gradual spread of the disease northward, from centres

where it had a strong hold in the adjacent counties southward, to the south bank of the Dee ; and, should that unfortunately take place, the utmost care and caution would be necessary on the part of those having stock on the north bank of that river. To make the boundary more definite, it was agreed by the Central Committee, on 12th January, that the parish of Banchory-Ternan be admitted into the Association.

The absence of any effective action in the counties immediately southward of Aberdeen has been already hinted at. It is not our purpose here to refer in any comprehensive way to the state of matters in Kincardine-shire and Forfarshire. Some general reference thereto seems, however, necessary in the way of contrast ; and as showing how serious a cause of anxiety the state of matters in these counties continued to be to those who, while fully alive to the contagious character of the disease, were convinced that it could not be successfully treated ; who held that there was no effectual remedy but that of "stamping out" by slaughtering diseased animals and such as had been in contact with them, all reasonable freedom being, on the other hand, conceded where due precautions had been taken to guard against the introduction of the plague.

In a report by the Honorary Secretary on the later outbreaks of the disease in Aberdeenshire and their origin, dated 14th February, 1866 (four fresh cases having occurred between 24th January and 9th February), it was stated that "since the 1st of January about sixty farms in the parish of Fettercairn and immediate neighbourhood have been attacked by the disease, and, as the farmers have attempted to cure, that district necessarily became a hot-bed of *rinderpest.*"

The endeavour, it may be stated, had been to find out whether these fresh outbreaks (with an intermediate one that had occurred in the parish of Strachan) were due to atmospheric agency—*i.e.*, the wind carrying infection from Fettercairn, in a north-easterly direction. Again, in a letter dated Ellon, March 1, 1866, on "the alleged cures of *rinderpest* in Kincardineshire," Mr. Thomas Hay, V.S., one of the most skilled and courageous of the Aberdeenshire County Inspectors, reported having, along with Mr. James Thomson, V.S., Aberdeen, at the request of the Honorary Secretary, visited a number of farms in Laurencekirk and Fordoun districts, where the disease was said to prevail. The Fordoun Inspector, who, according to Mr. Hay, was believed to be able to detect *rinderpest* "fourteen or twenty days before such an observer" as himself could discover any symptoms of the disease, accompanied the two Aberdeenshire Inspectors in their visit to several herds said to be labouring under *rinderpest*, but where the visitors, on examination, failed to see the least symptom of the plague! This Inspector had, moreover, it was caustically added, discovered an unfailing specific for *rinderpest* in the shape of cooked food, such as "boiled turnips and bran," as also "oatmeal gruel, salt being mixed in everything they were offered," and the supply kept short. It was found as they proceeded that while serious cases of plague had occurred, no care had apparently been taken to prevent the spread of the disease. At several farms, animals were found in the last stage of *rinderpest*, and a good many had died or been killed. In a few instances the complaint was found to be only foot-and-mouth disease, and at one reputed centre of plague, where there was a fine stock of cattle,

Mr. Hay's report was that " there is nothing apparently the matter with them but hunger, good indications of which they gave, by continuous lowing. Like all the rest, they are not allowed anything but a short supply of cooked food." The general conclusion was that the delusion—as delusion it certainly was, if allowed to go on—as to fancied cases of *rinderpest*, and its fancied cure, must be very serious, while " the recklessness of going from places where disease actually exists to places where it exists only in imagination, must be ruinous." Up to this date 101 cattle were reported to have died in Kincardineshire, and only 28 been killed. Thus far of Kincardineshire. The contrast between the course of procedure there, as graphically described in the letter quoted from, and the competent and fearless action so promptly adopted in Aberdeenshire to master the terrible scourge, is certainly sufficiently marked.

In Forfarshire (in parts of which plague was thereafter reported to have broken out afresh, with much violence, so late as the month of May) the loss was very heavy. The Local Authority, which sat with closed doors, had, from the outset, shown no disposition to carry out the provisions of the Act ; or if so, there had been no unity of action in devising or carrying out practical measures. And thus the state of matters there too formed a source of continued anxiety. Nor was the action taken in the county of Perth much more satisfactory, the number of cattle that had died of plague having been 108 up to the close of March, and the number killed only 8. At a general meeting of the Association, on 13th April, the laxity of the Local Authorities in the two counties first named, as well as elsewhere, in failing to carry out the existing law by slaughtering diseased animals, was somewhat severely commented upon ; and, at the meeting of the Central Committee, held directly after, it was resolved that this Association memorialise the Government to put the killing clauses of the Act everywhere in force ; and that the slaughter of diseased animals should be continued until the disease was exterminated. At the succeeding meeting, on 20th April, it was reported that the Home Secretary, Sir George Grey, had been communicated with in terms of this resolution, the appointment of a Commissioner to see that the Act was properly carried out being suggested. By returns for the week ending 7th April, disease was reported to exist only in nine counties in Scotland ; and, excepting in Forfarshire and Kincardineshire, no animal had been allowed to die. In the former, out of 58 attacked, 28 had been allowed to die ; in Kincardineshire only one had been allowed to die, giving evidence that the Act was now being more rigorously enforced in that county.

In Aberdeenshire it was reported that there was no disease whatever in the county at this date, the last animal affected having been slaughtered on 3rd April. There being no disease and no prospect of business, the Central Committee now agreed to adjourn for a fortnight, thus bringing their weekly meetings, which had been continued for more than seven months, to a close. There was no further outbreak of disease, and consequently no renewed necessity for resuming the stated meetings.   -

Of other steps taken by the Central Committee with a view to secure general action on the same lines as had been followed by the Association it is hardly necessary to speak.   But in drawing our observations on this

period of the Association's experience to a close, it may
be worth while to cite the terms of, practically, the last
regular report of the state of disease in the county, given
in by the Honorary Secretary at the meeting of the
Central Committee held on 2nd March, which brings
out distinctly that the Association had seen no cause to
doubt the soundness of the principles on which it had
acted throughout. After stating that no new cases had
been reported for the week, it went on :—

"In several of the later cases we have endeavoured
to restrict the slaughtering as much as possible, to avoid
what might be called indiscriminate slaughter, with the
view of saving, if possible, part of a stock, but, I regret
to say, the results are disappointing. At Banchory,
Mindurno, Oldfold, and Hill of Fiddes, the cattle in
separate byres were not slaughtered till disease appeared
amongst them, but of the whole only four young cattle
at Oldfold remain. These we may now begin to hope
will escape.

"From our late experience, therefore, we may con-
clude that if an animal is allowed to die of *rinderpest*,
there is very little hope of being able to save any
animal about that farmstead.

"I therefore urge upon farmers the necessity of
exercising the utmost vigilance in detecting the first
symptoms of disease, and these are very characteristic.
The animal attacked first refuses turnips, but continues
to eat straw for a day, or perhaps two days. This
arises probably from the soreness of the gums in the
front part of the mouth, which makes it painful to eat
turnips, although straw may be chewed when it has got
into the back part of the jaws. When the least suspicion
is awakened, let the animal be separated, and a com-

petent Veterinary Surgeon at once called in.   The sacri-
fice of only one animal may save the whole of the herd."*

The total expenditure incurred by the Association for
the three months to 31st March was £1136 13s. 1d.
The additional claims on the funds by sufferers from
rinderpest amounted to £829 17s. 1d., and, of the balance,
the item of watching during two months amounted to
no less than £214 10s. 1d.   Three cattle had died of the
value of £29 15s.; 51 had been killed and buried, of the
value of £530 10s.; and 89 had been killed and sold,
of the value of £1491 4s., the proceeds from sales of
carcases of the latter being £860 0s. 7d., which left a
loss of £631 3s. 5d., and the gross loss amounted to
£1042 18s. 10d.   The total claims for each of the
parishes in which disease had occurred during the three
months was as follows :—

Skene	£21	19	5
Old Deer ...	141	4	8
Banchory-Ternan	79	19	5
Oldmachar ...	195	17	0
Peterculter ...	120	18	10
Foveran	250	0	0
Longside ...	19	17	9
	£829	17	1

At this date the Association had still a free balance
on hand, after meeting all charges, of £703 6s. 4½d.

With the exception of a single animal killed by order
of the Inspectors, no fresh claim thereafter came on the
funds of the Association.   Claims of considerable

* Appendix B.

amount had been put forward by two members of the
Association at Middle Hythie, Old Deer, and Berryhill,
Peterhead, respectively ; but in each case the claim was
found to have been forfeited by the claimants having
brought cattle into the county contrary to rule IX., and
payment had been resisted accordingly.   In their report
to the general meeting, held on 20th July, 1866, however,
the Central Committee stated, *inter alia*, that they had
resolved " that, whilst the disease existed in the county,
it was absolutely necessary to rigidly enforce the articles
of Association ; but now that the county has been for
three months happily free of disease, the Committee are
disposed to act leniently toward both of these sufferers,
whose losses have been very considerable ; " and they
proposed—the proposal having been carried by a
majority in the Committee—to make an allowance of
£100 to each of them.   This was accordingly done ; and
though the question of a renewed assessment for the
succeeding year had been under discussion at a previous
meeting, it fortunately did not become necessary to
make any fresh levy, the final statement, submitted to a
meeting of the Rinderpest Association, on 19th October,
1866, showing a free balance in favour of the Association
of £428 3s. 5½d.

VI.—*General Summary.*

Reference has already been made to the statement
drawn up by the Honorary Secretary of the Rinderpest
Association, and laid before the meeting of the Central
Committee, for whom it had been prepared, on 15th
December, 1865. This statement, in accordance with

the wish of the Central Committee, was, with relative
resolutions, transmitted "to the Privy Council, Royal
Commission on Cattle Plague, and others interested
throughout the country." It was also in its completed
form submitted to the General Meeting of the Association held on 12th January, 1866, as a " Report on the
Origin and Progress of Cattle Plague in Aberdeenshire,
the proceedings of the Association, and the measures
best adapted to extirpate Cattle Plague." That the
report was of very great practical interest and value,
was generally admitted. And one marked illustration
of the attention excited by the proceedings of the
Association, as therein recorded, was furnished by a
communication, received from the then Chancellor of
the Exchequer, Mr. Gladstone, by Mr. M'Combie,
Tillyfour, as chairman of the Central Committee, and
laid before the meeting of that committee, on 19th
January. It was dated, " Westmacott Terrace, London,
January 12th, 1866," and ran as follows :—

" The Chancellor of the Exchequer presents his com-
pliments to Mr. W. M'Combie, and begs of him the
favour of one or more copies of the valuable paper (a
folio sheet), sent to the Home Office ten or more days
ago, and signed by Mr. M'Combie.

" This paper has only been seen by him to-day ; he
first heard of the proceedings in Aberdeenshire, so
honourable to the intelligence and forethought of that
county, through an Aberdeenshire man named Storer,
lately appointed Inspector in Chester.

" Would Mr. M'Combie be so kind as to have copies of
the papers sent to Robertson Gladstone, Esq., Liverpool,
and to Gregory Burnett, Esq., Dee Cottage, Flint. If

this statement has been printed in any public journals to which reference can be given, a reference to those periodicals would stand in lieu of the foregoing request.

"The Chancellor of the Exchequer's apology for taking this liberty, will, he hopes, be found in his great desire to render the Aberdeenshire proceedings extensively known."

The Paper, prepared by the Honorary Secretary, in its narrative part, gives an outline of the general proceedings of the Association, at once so concise and comprehensive, that, even at the risk of a little repetition, we make full quotations :—

"*Organisation.*—A public meeting of the county was held on 18th August last, at which the Lord-Lieutenant presided, when it was resolved to form this Association, with a Central Committee, supplemented by Parish Committees, to conduct its business. Funds were provided by a voluntary assessment of one penny per pound on the agricultural rental of proprietors and the same on tenants. The whole of the eighty-four parishes in the county joined the Association, and the assessment has yielded £4000, which is about four-fifths of the maximum realisable. Sufferers by disease are indemnified to the extent of two-thirds of their loss when the diseased animals are allowed to die, and three-fourths when slaughtered.

"*Procedure.*—The Central Committee, after a brief experience of the disease, came to the conclusion that "slaughtering out" was the wisest policy, and issued general instructions to the Parish Committees to the following effect:—' If disease was discovered in its earliest stages, before secretions at the eyes had taken place, only the animal affected to be slaughtered, the remainder of the herd being narrowly watched, and if disease again appeared, the whole animals in contact to be slaughtered. If the disease was not discovered till fully developed, then the whole animals in contact to be slaughtered at once.' These instructions are carried out by the Parish Committees, which meet as soon as a case of disease is reported, value the whole stock on the farm, confer with the Inspector, and then resolve what ought to be done in the special circumstances of each case. A copy of their resolution is submitted to the owner of the diseased animals, and if he fail to comply with the recommendation of the Committee, he forfeits all claim on the funds.

"The recommendations of the Parish Committees have in all cases been willingly agreed to by the members of the Association.

"The Parish Committees are also charged with the duty of seeing that the animals, &c., are properly buried, the byres disinfected, and all precautions used. It will easily be understood how much their own interest stimulates their zeal in attending to these matters.

"*Progress of the Disease.*—The disease had up to 10th January been brought into the county at various times to seven different places, which I shall call centres of disease.

"The *First* case of disease appeared at Bogenjohn, parish of Strichen, having been brought by a calf from London, received about 20th June. The first animal died on 1st July, and twenty-one died at intervals up to 4th August, when the farmer sent off for slaughter the.

remainder of his herd (twenty animals) which had been in contact with disease. Part of his stock kept on a distant part of his farm was reserved, and has not been attacked. From this farm the disease spread to Strichen Mains, about a mile distant. Here the first animal died on 11th August. Nine died at intervals up to 9th September, when the remainder (excepting three calves which had recovered) were slaughtered, and the plague was arrested.

" The *Second* centre of disease was at Mr. Ogg's, Westerton, parish of Forgue, to which the disease was brought by a calf from London, received on 18th July. Here the first animal died on 25th July, other 18 died up to 17th August, when the rest were slaughtered. After an interval of twenty days, the disease broke out at an adjoining farm (Mrs. Booth's), but the animals were immediately slaughtered (8th September), and there has been no case of disease in that district since.

" The *Third* centre of disease was at Garmond, parish of Monquhitter, brought by a calf from London, received on 18th July. The first animal died on 1st August. The disease spread to other three small crofts ; but as soon as it was ascertained to be *rinderpest*, all the cattle which had been in contact were destroyed on 16th September, and there has been no case since.

" The *Fourth* centre of disease was at Berryhill, parish of Peterhead, brought to this place also by a calf from London, received about 18th July. In this case the disease had spread considerably before proper measures were adopted, and was carried by cattle which had been grazing in adjacent fields to the parishes of Longside and St. Fergus. In Peterhead the disease appeared among six different herds, but was put down by

slaughtering on 28th September, since which time there
has been no disease in that parish.   In St. Fergus the
first animal died on 16th September, and eight different
herds were attacked, but the disease was exterminated
by slaughtering, and there has been no case since 7th
October.   In Longside the disease first appeared about
25th September, and two herds were attacked.   The
disease was put down for the time on 19th October.
After an interval of twenty-five days, a new outbreak
occurred on 13th November, supposed to proceed from
the fifth centre (to be afterwards mentioned).   In the
second outbreak, three herds were attacked, and the
disease was put down on 28th November, since which
time there has been no case in that parish.

"The *Fifth* centre was at Middle Hythie, situated in a
part of the parish of Old Deer, which runs into the
parishes of Longside and Lonmay.   To this farm the
Plague was brought by a lot of cattle from the south of
Scotland.   The cattle arrived on 11th October, and were
inspected on 15th and 18th, when they appeared healthy.
The disease, however, broke out with great virulence a
few days after.   The owner having forfeited his claim
on the funds of this Association, by bringing cattle into
the County contrary to its rules, kept the disease secret,
and it was not discovered by the Parish Committee till
eight or nine had died.   Part of the herd was driven off
to Peterhead alive and sold ; but the authorities inter-
fering, the rest, with the exception of a few which had
been at a distance from the others, were slaughtered.
From Middle Hythie the disease spread into the parish
of Lonmay, where three different stocks were affected
during the week from 9th to 13th November ; but the
cattle being promptly slaughtered, the disease was put

down for the time in that parish. The disease next appeared at Nether Hythie, where a cow standing in a byre with several others was observed to be ill, and was slaughtered on 24th November. This appeared to have stopped the disease, and there was no other case in that district for twenty days, when it again broke out simultaneously at three different farms within 500 yards of Middle Hythie, the original centre, and at a fourth farm distant about a mile. A cow at Nether Hythie, standing in the byre behind where the one previously slaughtered stood, was observed to be unwell on 13th December, and at each of two other farms within a day or two of the same time. Immediate action was taken, and the animals slaughtered as quickly as advisable, the last being killed on the 26th December. There has been no disease in the county since that date. The last outbreak at a new farm was, as has been stated, about 15th December.

"The *Sixth* centre of disease was at Stranduff, parish of Kincardine O'Neil (about forty miles from any of the previous cases). The disease was brought there, on 27th November, by three beasts, which had formed part of a lot brought over the hills from Forfarshire six days previously. The first animal died on 29th November, and the second on 7th December ; not till then did the owner suspect the disease to be *rinderpest*. Several animals died on the 9th, and the rest of the herd (in all twenty-four) were killed on 11th December.

"The *Seventh* centre of disease was at Auchinclech, parish of Skene, brought there by nine beasts belonging to the same dealer who sold the three to Stranduff. These animals were taken to Auchinclech on 5th December, first showed symptoms of disease on 10th, two died on the 13th, and the remaining seven were slaughtered the same day. Two cows which were for two or three days after the arrival of the diseased beasts in an adjoining byre, but which were removed a distance of 200 or 300 yards, before disease was apparent, showed symptoms on 18th December, and were immediately slaughtered. The disease has not spread from the sixth or seventh centres, which is rather remarkable, for both lots of cattle were driven several miles through the county, and the animal which died at Stranduff two days after arrival must have been travelled in an advanced stage of the disease."

At date of the statement the disease had been put down at the first four centres, and the places to which it had spread from these. At the fifth centre there had been no outbreak from 15th December, and at the sixth and seventh centres the disease had been put down immediately on discovery. In the subsequent report by the Central Committee, submitted to a public meeting of the Association, on 13th April, the following further statement was made, which may here be given to complete the record :—

"The county continued free of Cattle Plague till 19th January, on which day the disease broke out simultaneously at New Banchory and Pitmillan. On 23rd January it appeared at Mindurno ; on 1st February at Oldfold ; on 9th February at Millbrex ; on 11th February at Hill of Fiddes ; and on 24th February at Millhill. These are all the herds which have been attacked during the last three months. With the exception of the last (Millhill), all the farms are several miles apart from each other, and from any farm where disease has existed, the shortest distance being three miles. In the case of Millhill, there is reason to believe that infection was in some way communicated from Millbrex—distant about one mile.

"In all previous instances where disease appeared in this county at a new centre, it was clearly proved that it had been introduced by cattle which were themselves the first victims of the Plague, but as regards the preceding six first-named cases, it is as clearly established that the disease was not brought to any of the farms by cattle. Neither did particular inquiry elicit any fact warranting suspicion that the disease had been communicated by contact with individuals or articles which might have conveyed infection.

"Referring to the atmospheric phenomena of the period in question, observations recorded at Aberdeen show that on 14th, 18th, 28th January, and 14th February, hurricanes of wind from S.W., without any rain, swept over Aberdeenshire ; and it is to be noted that in four of the cases disease became manifest on the fifth day after a hurricane, in one case on the fourth day, and in one case on the seventh day. Since the 4th February there has been only one gale of wind from S.W., and it was accompanied by rain or sleet.

" Further, the six farms lie between north and north-east from the parish of Fettercairn in Kincardineshire, where disease was very prevalent at that time. New Banchory, the nearest of the farms to Fettercairn, is distant from it about 15 miles on the map ; and Millbrex, the most distant, about 40 miles.

"If the inference apparently deducible from these facts be correct, it is evident that the proceedings in reference to Cattle Plague, adopted in neighbouring counties, are of very great importance to us, and in consequence the Central Committee has at various times urged on the Government the necessity of taking measures to provide for the due and efficient discharge of their duties by Local Authorities under the 'Cattle Diseases Prevention Act, 1866.'

"As soon as disease was discovered at the various places, the animals affected, and, in some instances, all in contact, were immediately slaughtered ; and our action has been attended with the best results, for the disease has not spread, so far as discovered, from any of the centres, except in the case of Millhill, already alluded to.

"At Millhill, 19 animals have been slaughtered on account of the disease, and 44 still remain unaffected. The last case occurred on 3rd inst., when one animal was found to be affected after an interval of fourteen days, during which no disease had been visible about the farm. It is hoped the remaining forty-four animals will escape being attacked."

The hope here expressed was fulfilled ; the " plague," as already stated, having been finally "stayed" by the last case of slaughter recorded in the immediately preceding paragraph.

In the subsequent portion of the report by the Honorary Secretary, the following cases were noted in detail as particularly interesting and instructive :—

"*Case* 1.—The disease in the parish of Strichen where only two herds were attacked, was communicated from Mr. Keith's to Mr. Baird's stock, although it was clearly proved that there had. been no contact between any of the animals, neither had any of Mr. Baird's stock, subsequent to disease breaking out been on the same ground as any of Mr. Keith's. The disease existed at these two farms from 1st July to 9th September, and must have been communicated from the one herd to the other otherwise than by the movement of cattle.

" *Case* 2.—In Forgue the disease attacked only two herds, The last of Mr. Ogg's animals was slaughtered on 17th August, and Mrs. Booth's cow was not attacked till 6th September—an interval of twenty days. The distance between the farms is about 200 yards, but there had been no contact, and Mrs. Booth's cows were not upon the same ground as Mr. Ogg's subsequent, or even for some time previous to, disease breaking out. In this case disease was not caused by the movement of cattle.

" *Case* 3.—On 23rd November, a cow at Nether Hythie, within 300 yards of Middle Hythie, the centre of disease, was observed to be unwell, in consequence of loss of appetite, and the supply of milk failing. The parish and county Inspectors examined the cow on 24th. There was no external symptom of disease, no tears from the eyes, but the vagina showed the characteristic appearances of *rinderpest*, and the fourth stomach, on a *post-mortem* examination, clearly confirmed the existence of Cattle Plague. The cow was immediately slaughtered (24th November), and the byre cleansed and disinfected, but no preventive medicine was administered to the other cows. No further case of disease occurred for twenty days, when on 13th December

another cow (standing behind the one previously
slaughtered) was observed to be unwell and was
killed the day following. Up to this date the remaining cows in the same byre have not been
attacked. It is doubtful whether the second cow
got the disease from the one first killed, or whether
it was a fresh communication from the original
centre (distant some 300 yards only). It is remarkable that a calf at Mr. Mitchell's, another at Mr.
Robertson's, and the second cow at Nether Hythie,
three places all within 500 yards of the original
centre, were affected within a day of each other,
after an interval of twenty days, during which there
had been no disease in the neighbourhood. It may
be surmised that contagion had, by atmospheric
influence, been conveyed from manure, or offals
of deceased animals imperfectly covered at Middle
Hythie, where 28 animals had either died or been
slaughtered in an advanced stage of disease, produc-
ing a large mass of contagious matter.

" *Case* 4—Disease broke out at Greenwards in a byre
containing five cows. One showed external symptoms, and some of the others were slightly affected.
The five cows were slaughtered on 16th November,
and the rest of the stock (22 animals), all in adjoining houses, still remain unaffected.

" *Case* 5.—In contrast with the two preceding cases is
that at Longmuir. There a calf was found to be
unwell, and was treated by an unskilful practitioner,
who assured the owner that it was not labouring
under *rinderpest*. As a precautionary measure, the
calf was separated from all other animals, and
treated for some days, when it died. The In-

spectors were called in, and found that the calf had died of *rinderpest.* By this time nearly all the owner's stock, consisting of twenty-seven animals, were more or less affected, and the whole had to be slaughtered."

The Parish Committees had been instructed to enquire into the origin of each new outbreak of disease, and in order to ascertain as closely as possible how disease had been communicated from one herd to another, queries had been sent to the committees of seven parishes by the Honorary Secretary, and these, with abstract of the replies, were as follows :—

" *Query* 1.—How many herds have been attacked in your parish, exclusive of the original centre?
" *Answer.*—In all the parishes 28 different herds.
" *Query* 2.—In how many of these cases was the Cattle Plague communicated from one herd to another by direct contact of cattle, or by the animals being on the same ground, at different times?
" *Answer.*—In two cases known to be by contact.
" *Query* 3.—In how many cases are you certain that disease was not communicated in either of the ways stated in the preceding query?
" *Answer.*—Cannot be absolutely certain, but could not discover how it was possible that the disease could have been so communicated in 23 cases.
" *Query* 4.—How many cases are doubtful?
" *Answer.*—Three.
" *Query* 5.—Judging from your experience, could a total stoppage of the movement of cattle be depended on to put an end to the Cattle Plague?

"*Answer.*—Six answer no ; one believes it would, but does not think it practicable.

"*Query* 6.—Are other measures, such as those adopted by this Association, equally necessary ?

"*Answer.*—Six Committees reply quite as necessary ; one answers undoubtedly as necessary, but I should think not so certain.

"*Query* 7.—Would it be sufficient, as far as the movement of cattle is concerned, to prevent any movement of animals within a certain distance of any affected animal or farmstead, this being in addition to the prohibitions already in force as to bringing cattle into the county, &c. ?

"*Answer.*—All think it would be sufficient.

"*Query* 8.—To what distance from disease should the movement of animals be stopped ?

"*Answer.*—Three say half-a-mile ; three, one mile ; and one, two miles.

"Mr. Hay, V.S., the Inspector for the county, replied to the same queries to the following effect :—' Has attended officially twenty-three herds, exclusive of original centres. Of these, contagion was communicated in one case by direct contact ; in two cases, by being on the same ground or eating the same grass ; in seven cases, supposed by men ; and in thirteen cases, the medium is not even conjectured—but pretty certainly not by the contact or movement of cattle in all the cases except one, which is doubtful. Most emphatically the stoppage of the movement of cattle would not stop the Plague. Within a radius of one mile from Middle Hythie, no cattle have been moved from their stalls since 4th November, and round that centre there have been

thirteen different outbreaks since that date. Believes it quite sufficient to stop the movement of
cattle in the neighbourhood of disease. Under the
management of this Association, where seldom more
than one animal is allowed to die, one mile would
be sufficient ; but where treatment is resorted to,
and a great number die, as in the case of Middle
Hythie, three miles of a radius might be required.' "

Systematic observation and inquiry as to the orgi-
nation of the disease cannot be said to have been carried
further than is shown in the extracts given ; and for the
very sufficient reason that the *rinderpest* had been extirpated in the county. It only remains to say that,
while the successive Orders issued by the Privy Council
for dealing with the Cattle Plague were to an appreci-
able extent shaped in accordance with the expressed
findings of the Association—which declined to approve
of the extreme step of prohibiting all movement of cattle
during the prevalence of Plague, recommended by a
majority of the Special Royal Commission, as the one
effectual remedy—the policy of "stamping out" by
slaughtering, with part compensation to owners, was
more and more seen to be the sound and safe course of
procedure. And that principle was embodied in the Bill
for the Suppression of Cattle Plague introduced into the
House of Commons in February, 1866, by the Home
Secretary, Sir George Grey, and thereafter passed into
law.*

Abundant testimony to the success of the Aberdeenshire method of " stamping out " might be quoted from
the columns of leading London and other journals of the

* Appendix C.

time ; but there is no need to swell this brief record by doing so.   As tersely expressed by an able writer in one of these journals, in commenting upon the statement of the Honorary Secretary, the moral which the example set had for owners of stock was this — " Trust to yourselves in the first instance, and don't delay.   Organise, assess, offer compensation.   Then take the poleaxe in hand and kill without asking many questions.   That is what Aberdeenshire has done ; and hence, while of all counties in the Kingdom it had most to dread from the *rinderpest*, it is perhaps the one that will feel the effects least."   And very much in the line of practical enforcement of what is here laid down, as well as of commendation of the intelligence and fearless promptitude exhibited by the Executive of the Association were the words of Mr. Gladstone, when on 26th September, 1871, as Liberal Premier of the time he received the freedom of the city of Aberdeen.   After references to recent political movements in the county of Aberdeen, the Premier said :—

" But there was another service, and a marked service, that Aberdeenshire did to the country at that very period, in the winter which separated the years of 1865 and 1866.   I allude to the cattle plague, and I wish to say here that which I have said elsewhere in public and in private, that it was an admirable spectacle, when all over the country we were wandering and groping about, some proposing the most absurd measures by way of remedy and precaution, and others feeling themselves to be totally in the dark—it was an admirable spectacle when the gentry and farmers of the county of Aberdeen, associating themselves together, with nothing to rely upon except their own energy, except their own prudence and intelligence, devised for the ready, rapid, and complete extinction of that plague, the very remedy which, at a later period, after much ineffectual discussion, the Legislature found itself counselled by prudence to adopt. I cannot recollect, my Lord Provost, so remark able an example of local activity, self-reliance, practical ability, and wisdom, holding up for the whole nation a standard which that nation was ultimately glad to follow. And now, if ever that disease should unfortunately appear among us again, we have only got to put in operation your remedy—the remedy by which you, of the county of Aberdeen, taught us, with full assurance, and with the blessing of Providence, the mischief would be brought to a speedy and complete termination."

With these emphatic words we may fitly close this account of the Constitution and practical procedure of the Aberdeenshire Rinderpest Association.

APPENDICES.

APPENDIX A.

CONSTITUTION, RULES, AND REGULATIONS OF THE ABERDEEN-SHIRE RINDERPEST ASSOCIATION.

CONSTITUTION.

I. The Association shall be known as the "Aberdeenshire Rin-derpest Association," of which all Proprietors and Tenants shall be members for the year ending 30th June, 1866, who pay an Assessment of One Penny per Pound on the agricultural rental of their Estates or Farms on or before 30th December, 1865, or who, not being liable to that Assessment, otherwise contribute to the funds of the Association such sum as may be fixed by the Central Committee. But the Central Committee shall have power to admit members after the date aforesaid.

II. No Member shall, under any circumstances, be liable for any amount beyond the sum subscribed by him.

III. The sphere of the Association shall comprehend the whole County of Aberdeen, together with the part of Banffshire situated within it. And the Central Committee shall have power to include, if it seem advisable to them, the part of Kincardineshire which lies north of the Dee, and the Parish of Inverkeithing in Banffshire.

IV. The objects of the Association are—1st, The extermination within the County of the Cattle Disease, commonly known as the *Rinderpest* or Cattle Plague ; and 2nd, The indemnification of Sufferers by this Disease, to the extent and subject to the Rules and Conditions after-mentioned, or such Rules and Regulations as may from time to time be issued by the Central Committee.

CENTRAL COMMITTEE.

V. The whole powers and funds of the Association shall be vested in a Committee to be called the "Central Committee," which shall have full powers to indemnify sufferers by the disease as under-mentioned, to delegate powers to Parish Committees, and generally to take such measures as may to them seem advisable for accomplishing the objects of the Association. All payments indemnifying sufferers shall be voted by the Central Committee.

VI. The Central Committee shall consist of the following gentlemen :—

> The Earl of Kintore ; Alexander Forbes Irvine of Drum, Convener of the County; Sir Alexander Anderson, Lord Provost of Aberdeen ; Sir James D. H. Elphinstone, Bart. of Logie-Elphinstone ; William Leslie of Warthill, M.P. ; James W. Barclay, Aberdeen ; James Hay Chalmers, Advocate, Aberdeen ; Alexander Campbell, Blairton, Hill of Mennie ; James Cochran, Little Haddo, Newburgh, Aberdeen ; Sylvester Campbell, Kinnellar, Blackburn ; Anthony Cruickshank, Aberdeen ; George Cruickshank, Comisty ; Robert Copland, Mill of Ardlethen, Ellon ; Patrick Davidson of Inchmarlo ; William Cosmo Gordon of Fyvie ; Thomas Innes of Learney ; Robert Kemp, Grain Merchant, Aberdeen ; John Ligertwood, Advocate, Aberdeen ; William M'Combie, Tillyfour, Whitehouse ; William S. Marr, Uppermill, Tarves ; James Martin, Butcher, Aberdeen ; George Milne of Kinaldie, Dyce ; H. L. L. Morison of Blair ; George Reid, Seedsman, Aberdeen ; James Stewart, Butcher, Aberdeen ; Robert Williamson, Bendaugh, Dyce ; Alexander Sim, Fawells, Keith-hall ; John Gordon of Parkhill ; Hugh F. Leslie of Powis ; John Ross of Tillycorthie ; D. R. L. Grant of Kingsford ; William M'Combie of Easter Skene ; A. F. Douglass, Haddo House ; William M'Comble, Editor, *Free Press;* C. A. Barclay, Aberdour House ; William Milne, Accountant, Aberdeen ; Provost Alexander, Peterhead ; John Logan, Lunderty, St. Fergus ; R. Macdonald, Cluny Castle ; Alexander Mitchell of Kincrnig.

Also the Convener of each Parish Committee. The Central Committee shall have power to add to its number, and five members shall form a *quorum*.

PARISH COMMITTEES.

VII. The Parish Committees shall collect the Assessments, and remit through their Conveners to the Secretary of the Central Committee, watch over the Interests of the Association in their respective parishes, reporting to the Central Committee and District Inspector any suspicious cases of disease, or any cattle which may have been brought into the parish from places out of the county, or regarding which suspicions may exist, and generally assist in carrying out the instructions and regulations of the Central Committee.

GENERAL RULES AND REGULATIONS.

VIII. As soon as the owner of any cattle observes any disease amongst his animals, which he knows or suspects to be *Rinderpest*, or any disease which he does not know, he shall immediately inform the District Inspector, and the Convener of the Parish Committee thereof, and carefully attend to and comply with the instructions he may receive from either of them ; and should the disease be declared to be *Rinderpest* by any of the Inspectors for the county, the owner of the diseased animals shall be bound to place his whole stock of Cattle under the direction and control of the Central Committee and its officers, to be dealt with as shall seem most advisable to them for the public good, and subject to the rules and regulations here laid down, and which may from time to time be hereafter laid down by the Central Committee.

IX. All the Members of the Association shall be bound to adhere to, and observe the whole of the rules, regulations, and precautions prescribed in the Orders issued by the Privy Council, or from time to time by the Central Committee, or contained in any Act of Parliament, and shall give all necessary assistance in carrying out the objects of the Association.

X. Members of the Association who may be sufferers by the disease of "*Rinderpest*" shall, as a general rule, be indemnified to the extent of three-fourths of their loss where the animals have been slaughtered or disposed of by the Central Committee, or by a Parish Committee authorised by them, but under special circumstances the Central Committee shall have full power to diminish the proportion of indemnification.

XI. Sufferers by this disease who are not members of this Association, or members who shall infringe any of the preceding rules and regulations, or the rules, regulations, or precautions which may be issued by the Central Committee, or under any Order in Council, or any Act of Parliament having reference to the *Rinderpest*, shall have no claim on, or right to the funds of the Association, but the Central Committee shall have power to treat with or indemnify such parties to such extent, and in such manner as they shall deem advisable for the public good.

XII. The Central Committee shall be bound to see that the provisions of any Orders in Council or Act of Parliament are carried out, and to recover the penalties for non-compliance therewith.

XIII. At the end of every three months, from the 1st October, 1865, the accounts of the Association shall be made up, and a General Meeting of the whole Members shall be held in Aberdeen, to instruct the Central Committee as to farther proceedings, and as to the disposal of any Balance of Funds which may then belong to the Association, and which Balance shall be disposed of as the majority of Members present at such meeting may direct.

XIV. All claims made on the Association and passed and admitted by the Central Committee shall be payable out of the funds of the Association, so far as sufficient for that purpose immediately after the Quarterly General Meeting of the Association.

XV. Nothing in these rules contained, and no contribution given by a member or other person, shall entitle any one, whether a member of the Association or not, to any legal claim on the funds of the Association, or on its members individually.   The whole fund shall be at the disposal and under the discretionary control of the Central Committee, who shall be responsible only to the Association itself for the moneys impressed into their hands and the disposal thereof; it being understood and agreed on that the indemnification provided for above shall not be claimable as a debt or obligation by any member.

APPENDIX B.

The most recent outbreak of cattle plague under British surveillance occurred in the island of Cyprus in the end of 1879 and beginning of 1880. The following paragraph, describing the results of *post mortem* examinations, is taken from the report of Dr. Heïdenstam, Civil Surgeon of Larnaca, who acted under authority of the High Commissioner as Chief Inspector of Cattle Diseases. As will be seen, the description of symptoms agrees substantially with what was found when the disease prevailed in Aberdeenshire :—

"I have examined the bodies of over one hundred animals after death, and have found, almost invariably, that the interior of the carcases presented the same appearance. The inside of the mouth and pharynx were of a darkish red colour. The tongue was flabby, and covered with a yellowish exudation. I saw nothing remarkable about the three first stomachs, but on the fourth I noticed several deep red blotches, and in some cases it was spotted with small ulcers, which forcibly reminded me of those so common in cases of catarrhal inflammation of the human stomach. I was surprised to find the small intestines, generally speaking, free from disease. The bronchial mucous membrane was frequently injected and covered with tough mucous. The lungs were congested and swollen, and their interlobular tissue was distended with air. The heart was relaxed and discoloured. The brain appeared to be unaffected, although it was more than usually injected with blood, and the meninges were of a reddish colour. I noticed no particular change about any of the other parts, except that the body was generally more red than is customary, as were also the urinary and generative organs."

APPENDIX C.

The following general statement on the subject of cattle plague in this country is taken, slightly condensed, from the supplement to *Chambers's Encyclopædia*, published in 1868 :—

"The British outbreak of 1865-1867, like its predecessors, undoubtedly came from Russia. The steamer *Tonning*, from Revel, brought 331 cattle and 330 sheep into Hull on 29th May, 1865. A portion of the cattle had come from the interior of Russia, where the plague then was, or recently had been ; the cargo was rapidly landed, and very hurriedly inspected. Nearly half of the cattle were distributed in various lots to butchers in Leeds, Derby, and Manchester, but, curiously, these do not appear to have left any contagion in their trail. One hundred and seventy-five came to London, remained from the Monday evening until Thursday's market in lairs at York Road, adjoining the cattle market. It is stated in a leader in the *Times* of 15th August, that *rinderpest* was seen in the metropolitan market as early as 12th June. Certain it is that more than one lot purchased on 19th June carried the disease to several dairies in and about London. Until 11th August, 1865, no restrictions whatever were put upon the removal of cattle. As early as 18th July, the pest was brought from London to Huntly by four calves ; subsequent outbreaks occurred in the same way. The stamping-out system was, however, early and rigidly enforced in Aberdeenshire, and eight distinct outbreaks of the plague were promptly got rid of.

" In Edinburgh, it appeared probably about 9th August, was brought from London by some low-priced foreign cows ; and in six weeks, about 800, or one half the dairy cows in Edinburgh, had died—200 having been buried in one trench. By the end of January, four-fifths of the dairy cows had perished, but Edinburgh was reported clear. In Glasgow, the first case occurred on 19th August, in a cow sent from Edinburgh. By 30th September, 432 cases were reported, and it continued to spread. By the middle of October, it was in Mr. Harvey's valuable stock of 800, of which

25 died in one night, and to save further loss, 50 healthy animals were in one day disposed of to the butcher. From Falkirk Trysts, as from Darnet, Norwich Hill, and other large English fairs, the disease was transmitted into fresh localities. From the autumn trysts, it was carried into Perthshire, Forfarshire, and Fifeshire. Diseased cattle passing along in railway trucks appear to have spread the contagion over the fields adjoining the line at Thornton, Fifeshire. Into West Lothian it was conveyed early in September by lambs from the Edinburgh market.

"The rapid spread of the insidious disorder may be gathered from the fact that, whilst during the week ending 24th June, 1865, there was only one outbreak at Mrs. Nicholl's dairy at Islington, and 30 animals affected, by 30th September there were 1702 farms, sheds, or other places in which the pest had appeared, and 13,263 animals had been attacked. Three months later, 8252 separate places had been visited, and 62,743 animals attacked. During six months, the aggregate of cattle attacked was 76,002. During the three months to 30th March, 13,443 farms and other premises had been infected, and 147,275 cattle attacked. In December, 1865, the fresh cases each week reached 9000 ; but in spite of remedial and preventive measures, of Orders in Council, and restrictions on the movement of stock, the number of weekly cases steadily increased to 15,706 in the third week of February. 'The Cattle Diseases Prevention Act' passed 20th February, 1866, and the advantages flowing from the restrictions thus tardily imposed on the trade in cattle. and the slaughter of diseased and infected animals were speedily apparent. In four weeks, the number of cases was reduced by one half. During the three months ending 30th June, 28,276 cases were reported ; during the next three months to 30th September, the numbers fell to 2108 ; whilst to 29th December the three months' cases were but 149 ; to 30th March, 1867, 89 new cases were noted. Throughout April and May the number of cases continued steadily to decline ; but during the week ending 25th May a fresh outbreak occurred in the Finsbury district of the metropolis, and 81 animals died, or were slaughtered to prevent the further spread of the pest. With the exception of an isolated outbreak in Essex, which was promptly stayed by slaughter of the ailing and suspected animals, the country was free

of plague during August. By the 'Consolidated Order of Council'
(August, 1867), foreign cattle, from 13th September, 1867, are to be
slaughtered at the ports of debarkation, and the reintroduction of
cattle plague into Great Britain is thus tolerably effectually guarded
against.

"The total number of animals affected in Great Britain to 31st
August, 1867, has been 278,923 ; about 125,000 have died ; whilst
nearly 60,000 healthy cattle have been slaughtered to prevent the
further spread of the disease. The numbers of cattle in England,
Wales, and Scotland, attacked, killed, &c., may be thus approxi-
mately stated ; about 11,000 cases known to have been attacked
are, however, unaccounted for :—

	Attacked.	Killed.	Died.	Recovered.
England, . .	223,672	102,740	90,450	21,589
Wales, . . .	8,388	1,180	5,794	1,117
Scotland, . .	46,863	6,263	28,088	10,707
Total, . .	278,923	110,183	124,332	33,413

www.ingramcontent.com/pod-product-compliance
Lightning Source LLC
Chambersburg PA
CBHW021959190326
41519CB00010B/1330